MW00964532

VEHICLES ON THE JOB

FARM VEHICLES

BY JAMES BOW

NorwoodHouse Press

Cover: A farmer can use a tractor to pull equipment across the fields.

Norwood House Press
P.O. Box 316598
Chicago, Illinois 60631

For information regarding Norwood House Press, please visit our website at:
www.norwoodhousepress.com or call 866-565-2900.

PHOTO CREDITS: Cover: © CaseyHillPhoto/iStockphoto; © AS photo studio/Shutterstock Images, 11; © Avalon_Studio/iStockphoto, 21; © Denton Rumsey/Shutterstock Images, 17; © Deposit Photos/Glow Images, 4–5; © Igor Zehl/CTK/AP Images, 18–19; © jondpatton/iStockphoto, 7; © SimplyCreativePhotography/iStockphoto, 14; © Steve Debenport/iStockphoto, 12–13; © Veremeev/iStockphoto, 8–9

LIBRARY OF CONGRESS CATALOGING-IN-PUBLICATION DATA

Names: Bow, James, author.
Title: Farm vehicles / by James Bow.
Description: Chicago, Illinois : Norwood House Press, [2018] | Series: Vehicles on the job | Includes bibliographical references and index.
Identifiers: LCCN 2018005355 (print) | LCCN 2018003246 (ebook) | ISBN 9781684042296 (ebook) | ISBN 9781599539454 (hardcover : alk. paper)
Subjects: LCSH: Agricultural machinery--Juvenile literature.
Classification: LCC S675.25 (print) | LCC S675.25 .B69 2018 (ebook) | DDC 631.3/7--dc23
LC record available at https://lccn.loc.gov/2018005355

312N—072018
Manufactured in the United States of America in North Mankato, Minnesota.

CONTENTS

Note: Words that are **bolded** in the text are defined in the glossary.

Farmers work long days in the field to help crops grow.

FARMERS IN ACTION

Smoke puffs from the tractor. Its engine roars. Its wheels bite into black soil. The farmer steers from inside the glass and steel **cab**.

The tractor pulls a large plow. The plow looks like a metal skeleton hand. Its fingers dig into the ground. Long, straight lines of black earth stretch behind it. The ground is ready to be planted.

The farmer waves to his friend across the road. She's driving a combine harvester. It rumbles as it follows a line of wheat. It chops the tall yellow wheat. It spits the seeds into a grain truck. The vehicle leaves behind a line of chopped stalks.

A truck passes on the road. It has a big tank and long arms with spray nozzles. It's off to spray another field with **fertilizer**. It's a smelly job, but for some farms it is an important one.

The farm is a busy place. Lots of work needs to be done. Luckily, the farmers have big machines to take on big jobs.

PARTS OF A TRACTOR

Modern tractors may have comforts like air conditioning and a quiet cab for the farmer.

THE HEART OF THE FARM

The tractor is the heart of many farms. The tractor pulls the equipment that helps the farmer do his or her job. Its engine is in the front, under a metal hood. The driver sits inside a cab behind it.

The engine turns large wheels in the back of the tractor. These grip the soil. The smaller wheels at the front steer.

Some tractor engines are almost as powerful as race cars. But tractors are made to move slowly. They use that power for strength instead of speed.

The John Deere 9630 is one of the biggest tractors in the world. It is over 22 feet (7 m) long and 10 feet (3 m) wide. It can pull up to 54,000 pounds (24,000 kg). That means this strong tractor could pull 13 cars!

A strong engine gives a tractor enough power to pull heavy loads.

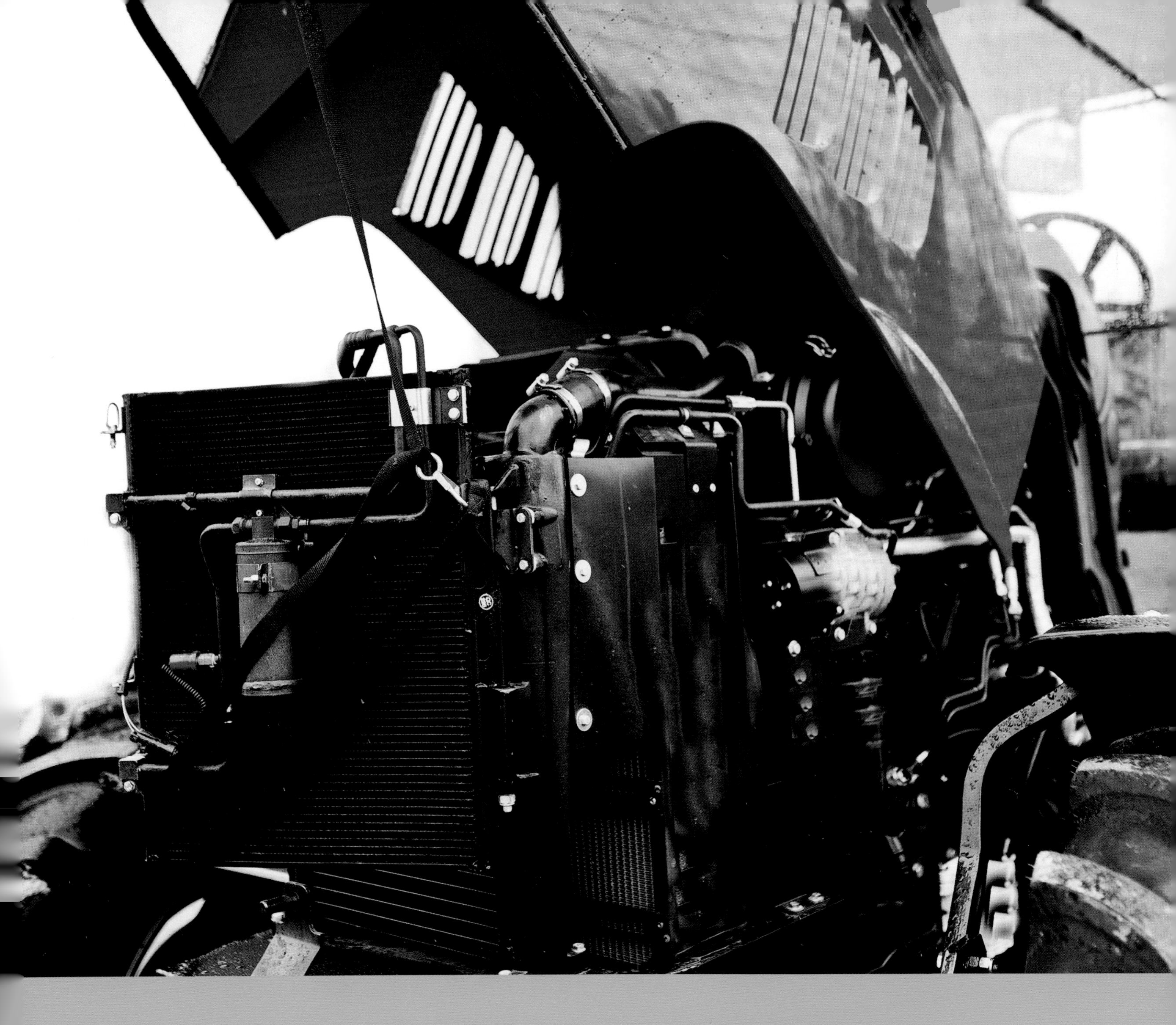

A tractor pulls a plow to prepare soil for planting.

CHAPTER 3

PREPARING THE SOIL

To grow crops, farmers need to plant seeds. The plow helps get the soil ready for planting.

Plows prepare the soil for planting by cutting into it and turning it over. A blade digs into the ground. Soil rises up from the front of the blade. Behind the blade, a moldboard takes this soil and folds it over. This creates a **furrow**, or groove, in the soil.

Plowing brings fresh **nutrients** to the surface. It also covers over weeds and the remains of old crops. As these old plants break down, they add more nutrients to the soil.

The plow is one of the oldest farming tools. The first plows were pushed by people. Later, animals like

A no-till seed drill is attached to a tractor for planting.

STEM AT WORK: MACHINE DESIGN

Engineers design machines to mimic the jobs farmers used to do. A stonepicker removes rocks from the field. A harrow breaks up clumps of soil. The plow has been carefully angled to cut through the soil more easily.

oxen and horses pulled them. When tractors started pulling them, plows got much bigger.

Early plows cut just one furrow. Today, they cut many at once. This gets fields plowed more quickly.

Plowing can also harm soil health. It affects soil structure. And, it can increase **erosion**. So, some farmers use other machines instead, such as no-till seed drills. These don't turn over the soil. They punch holes through what's left from old crops to plant new seeds.

CHAPTER 4

MAKING THINGS GROW

Long ago, farmers noticed that mixing animal waste into soil made their crops grow better. Scientists found the parts of the waste that helped the most. With this, they created fertilizer.

Spreading fertilizer is a stinky, messy job. Fertilizer used to be thrown out of wagons onto the field. Too much fertilizer can harm the crops. Farmers have to take care to spread out the fertilizer evenly.

Today, trucks with big tanks spray liquid fertilizer. These trucks have long arms that swing out. They spray large parts of the field all at once.

Farmers choose different types of irrigation based on their land and what crops they grow.

Farmers need water, too. If it doesn't rain, they have to **irrigate**. A farm might use 4,000 gallons (15,000 L) of water per acre per day just to make their crops grow. That's enough to fill about 100 bathtubs!

Getting water to the crops can be tricky. Many farmers can't run pipes through fields they hope to plow. One vehicle that can help is called a side roll or a wheel line. Long pipes are put on wheels that move along the plowed field. Water flows through these pipes and sprays the crops beneath.

A combine is a farm vehicle that helps harvest crops.

HARVESTING THE CROPS

After a growing season, it's time for the farmer to get the crops out of the fields. This is a difficult job. Potatoes and carrots have to be dug out of the earth. Wheat and corn stalks have to be cut down.

ON THE JOB: FARM SCIENCE

Farming involves a lot of science. Biologists study how crops grow. Veterinarians are doctors for animals on the farm. Engineers and mechanics design farm vehicles and keep them working. If you'd like to help on a farm, go to school to learn about the science of **agriculture**!

The parts we can eat must be separated from the parts we can't eat.

In the past, farmers did this by hand. Today, machines often do these jobs. Combine harvesters move around the fields. They chop down the crops.

A combine has different heads to harvest different crops. A grain head cuts down the stalks of wheat and moves them into an **auger**. The auger pulls the grain into the combine. Inside, a drum **threshes** the wheat and pulls away the grain. The grain pours out of a spout into a truck that drives alongside the combine.

A carrot harvester moves along the row of crops. Rubber belts grab the top of each carrot. The carrots are pulled out of the earth and cleaned. Then they are put into a truck driving alongside the harvester.

Machines like these help farmers grow more crops. Those crops feed people and animals. They provide material to make fuels. All this is possible because of farm vehicles.

Farm vehicles help farmers work hard all year round.

GLOSSARY

agriculture (AG-rih-kul-chuhr): The science of farming.

auger (AW-guhr): A spiral-shaped tool used for moving loose material such as grain.

cab (CAB): The covered area where the operator of a tractor or other machine can sit or stand.

erosion (ih-ROH-shun): When soil, rocks, and other landforms wear down and disappear.

fertilizer (FUHR-tuh-leye-zuhr): Something used to help crops grow, usually a chemical made from waste products.

furrow (FUHR-oh): A long groove in the ground made by a plow for planting.

irrigate (IR-uh-gate): To add water to help things grow.

nutrients (NOO-tree-uhnts): Substances that help living things grow.

threshes (THRESH-ez): Separates the grain seeds from the rest of the plant.

FOR MORE INFORMATION

BOOKS

Alexander, Heather. *Big Book of Tractors*. New York, NY: Parachute Press, 2007. Learn all about tractors, their history, and the work they can do.

Peppas, Lynn. *Vehicles on the Farm*. New York, NY: Crabtree Publishing, 2011. Learn about the hardworking vehicles found on a farm.

Total Tractor! New York, NY: DK, 2015. Learn about tractors and other farm vehicles with vivid photographs.

WEBSITES

My American Farm

www.myamericanfarm.org

Kids can explore this site to learn about the food they eat and the farms that make it.

National Agriculture in the Classroom

www.agclassroom.org/student

Learn about agriculture and careers in farming through games, videos, and virtual tours.

INDEX

ABOUT THE AUTHOR

James Bow is the author of more than 40 educational books for children and young adults, a novelist, and a local columnist. He graduated from the University of Waterloo School of Urban and Regional Planning in 1991. Born in Toronto, he now lives in Kitchener, Ontario, Canada, with his author wife and his two daughters.